Alcohol
and the
Bible

Alcohol and the Bible

Howard H. Charles

HERALD PRESS
Scottdale, Pennsylvania
Kitchener, Ontario
1981

Except as otherwise indicated, Scripture quotations are from the Revised Standard Version of the Bible, copyrighted 1946, 1952, © 1971, 1973.

ALCOHOL AND THE BIBLE
Copyright © 1966, 1981 by Herald Press, Scottdale, Pa. 15683
 Published simultaneously in Canada by Herald Press,
 Kitchener, Ont. N2G 4M5
Library of Congress Catalog Card Number: 66-10970
International Standard Book Number: 0-8361-1941-X
Printed in the United States of America
Design: Alice B. Shetler

81 82 83 84 85 10 9 8 7 6 5 4 3

Contents

Introduction — 7

1. The Witness of the Old Testament — 9
2. The Teaching and Practice of Judaism — 14
3. New Testament Teaching and Practice — 18
4. New Testament Ethical Perspectives — 24
5. Conclusion — 32

Notes — 37
The Author — 40

Introduction

A brochure advertising a well-known modern magazine offers "a gentleman's guide to being a modern gentleman." One of the earmarks of this distinguished species is, we are told, that he knows a little about drinks. "Two parts orange juice, one part brandy, one part Cointreau. Mix together with just enough champagne to fill the glass, and you'll be drinking just what Georges serves his patrons at the Paris Ritz. The potion is called a Pick-Me-Up, and ... [he] highly recommends it—no matter what city you happen to be in." The reader is then invited to place a little check before the following statement: "I am a gentleman who enjoys a new drink every once in a while, and could use a magazine like ... that regularly comes up with ideas for making them."

This type of advertisement is typical of the powerful pressures constantly exerted through the press, radio, television, and billboards to create an acceptance of the creed that the way to happiness, health, and prestige is through drinking alcoholic beverages.

The liquor industry reputedly spends about 600 million dollars annually in advertising. These appeals are not without a measure of success. In 1976 the American public paid more than 30 billion dollars for alcoholic beverages. Since 1971 the per capita consumption of such drinks has been the highest since 1850 ranging from 2.63 to 2.69 gallons per person over thirteen years of age. Conservative estimates place the number of adult problem drinkers, including alcoholics, between 9.3 and 10 million. This is 7 percent of the 145 million adults over 18 years of age. In addition, there are an estimated 3.3 million problem drinkers among youth from ages fourteen to seventeen. This is 19 percent of the 17 million persons in this age-group.

The calculated cost to the United States of alcohol abuse in 1975 totaled some 40 billion dollars in lost production, medical expense, automobile accidents, crime, fire losses, and other social costs. Half of the traffic fatalities and one third of all traffic injuries are alcohol-related. Alcohol is a known contributor to certain types of cancer and to heart disease. One out of every 2,000 babies born in 1978 had an impairment caused by the mother's consumption of alcohol during pregnancy. Recent studies have shown that up to 40 percent of fatal industrial accidents, 69 percent of drownings, 83 percent of fire fatalities and 70 percent of fatal falls were alcohol-related.

The recital of figures could easily be extended.[1] But statistics do not tell the whole story. Headaches, heartaches, accidents, disease, poverty, cruelty, isolation, growing helplessness, and deepening hopelessness frequently follow in the wake of addiction to alcohol. These cannot be comprehended in cold figures. Statistics alone do not provide the ultimate grounds for an evaluation of the problem. The Christian needs to bring this matter, as every other experience, under the light of God's Word. What guidance does the Bible offer in shaping our attitude toward the use of alcoholic beverages?

As on many other issues the witness of the Bible has been variously understood. Its authority has been appealed to by both the "wets" and the "drys" for their mutually exclusive positions. Indeed, biblical statements can be found which seem to support opposite points of view.[2] Thus right principles of interpretation are important in our approach to the Bible.

In this discussion a historical survey will first be made of the practice and teaching regarding the use of alcoholic beverages in the Old Testament, Judaism, and the New Testament. Then an attempt will be made to lay hold of certain general but basic principles which may give guidance in finding a Christian solution to the problem. If the first part seems unduly long, a careful documentation of the biblical data seemed important to correct whatever false impressions we may have. This part of the study should not obscure the effort to develop some general positive guidelines in the latter part of the booklet.

1

The Witness of the Old Testament

From the earliest times the cultivation of the vine and the making of wine were an important part of Israelite life in Palestine. Viticulture was well established in the land before the Hebrews entered it (Numbers 13:23; Deuteronomy 6:11). An exiled Egyptian official who observed life in Syria-Palestine near the beginning of the second millennium BC remarked that there was "more wine than water."[3] Probably the scarcity of water and its frequent contamination were practical incentives to produce wine. Both the climate and the terrain of the country were well suited to growing grapes. The large place which this industry occupied in the life of the Israelites is well supported not only by the many references to vineyards and wine in the Old Testament but also by archaeology. The remains of many grape presses of various types have been discovered, especially in the lowlands of Judah. The land thus could aptly be described as one of "wine . . . and vineyards" (2 Kings 18:32).

Wine in Israelite life had both a secular and a religious use.

It was a beverage familiar to all classes. Even the young were not excluded (Lamentations 2:11 f.; Zechariah 9:17). Sometimes bread and wine seem to have constituted the entire meal (Judges 19:19; Ruth 2:14). In other instances, wine along with bread and meat appear to represent the basic elements of nourishment (1 Samuel 10:3; 16:20). Wine is listed also as an essential item in the more elaborate catalogs of dietary provisions (1 Samuel 25:18; 2 Samuel 16:1 f). Doubtless among the poor its use was limited by economic necessity. In more affluent and aristocratic circles it was drunk in large quantities. Cf. Amos 6:4-6. In fact, the Hebrew word for "banquet" or "feast" actually means "drinking."

In addition to domestic use wine was employed in cultic rites and celebrations. It was used in sacrifice among not only the Hebrews but also the Canaanites and the Babylonians. In Israelite practice wine normally was not used independently but along with other sacrificial gifts (Exodus 29:40; Leviticus 23:13). While the amount to be used was specified, the precise way in which it was to be used was left indefinite (Numbers 15:1-10). However, from later Jewish evidence now available to us it appears that, like the blood, it was "poured out at the foot of the altar."[4]

The wine used by the Israelites was made almost entirely from grapes. Whether it was common in Old Testament times as in later Roman times to dilute the wine with water is not clear. However, wine was frequently mixed with aromatic herbs and spices to add to its flavor (Song of Solomon 8:2; Proverbs 9:2, 5; 23:30). Such additions may have increased its intoxicating quality. Archaeology has shown that the Israelites did not drink beer, although this was a common beverage among the Philistines.[5]

The important place of wine in Hebrew life indicates that the basic attitude toward it was essentially positive. It was one of God's good gifts to His people "to gladden the heart of man" (Psalm 104:15). The patriarch Isaac asked that God might grant "plenty of grain and wine" to his son Jacob (Genesis 27:28). Later, when the prophets looked forward to the day of Israel's renewal beyond the time of judgment, they spoke of it as a period when "the mountains shall drip sweet wine, and all the hills shall flow with it" (Amos 9:13; cf. Joel 2:24; 3:18).

But the Hebrews were aware also of the potential evils of wine which arose from its misuse. Drunkenness is frequently alluded to throughout the Old Testament and its familiar accompaniments are more than once graphically described. The prophet Amos denounced the women of Samaria who in their thirst for drink compel their husbands to oppress the poor in order to secure money to buy wine (4:1). Habakkuk warned that "wine is treacherous" (2:5) and Hosea complained that "wine and new wine take away the understanding" (4:11). Isaiah pronounced woe on those "who rise early in the morning, that they may run after strong drink, who tarry late into the evening till wine inflames them!" (5:11; cf. v. 22). Nothing but tragedy awaits that people whose religious leaders, the priest and the prophet, "reel with ... strong drink" and "are confused with wine" (Isaiah 28:7).

The Book of Proverbs is particularly aware of the dangers of wine and repeatedly warns against its use. While wine may be appropriate for those who are perishing or who are in bitter distress, "it is not for kings to drink wine, or for rulers to desire strong drink; lest they drink and forget what has been decreed, and pervert the rights of all the afflicted" (31:4 f.). "Wine is a mocker, strong drink a brawler," and the man who is wise will not allow himself to be led astray by it (20:1). Drunkenness will surely lead to poverty (23:21). But this is only one of its many bitter fruits.

> Who has woe? Who has sorrow? Who has strife? Who has complaining? Who has wounds without cause? Who has redness of eyes? Those who tarry long over wine, those who go to try mixed wine.

The admonition, therefore, which follows is understandable:

> Do not look at wine when it is red, when it sparkles in the cup and goes down smoothly. At the last it bites like a serpent, and stings like an adder. Your eyes will see strange things, and your mind utter perverse things (23:29-33).

It should be noted, however, that these and similar warnings are directed against the evils of excessive drinking. Nowhere in the Old Testament is the principle of total abstinence laid down as a rule applicable to all.

But abstinence from the use of wine was not unknown in Israel. According to Leviticus 10:9 the priests while officiating at the shrine were forbidden to drink wine under the threat of death. This prohibition is repeated in Ezekiel 44:21. But in neither instance is the reason for it given. It may have been a precautionary measure to avoid irregularities in practice due to intoxication. It was not a prohibition peculiar to Israelite priests but was observed also in some pagan cults of the ancient world.

Another instance of abstinence was the Nazirite, who by a solemn vow "cut himself off from the normal ways of life by abstention from certain things, and so put himself at the disposal of the deity as a special instrument."[6] The dedication in some cases was temporary, in others lifelong. It may have been a consecration to temple service as was true of Samuel or to a ministry of warlike feats against the enemies of Israel as in the case of Samson. Whatever variations existed in the purpose of the vow and its duration, the Nazirite was forbidden to drink wine while bound by the vow (Numbers 6:2-4, 20; cf. Judges 13:5, 7).

The Rechabites were another group who refused to drink wine. We learn of them in Jeremiah 35. The origin of this group is obscure. Apparently they were a seminomadic people who lived in the Judean wilderness. They likely intended by their way of life to protest the corrupting influence of Canaanite life in Israel. They refused to live in houses, till the soil, plant vineyards, or drink wine. The Rechabites regarded such practices as an accommodation to Canaanite sedentary civilization and thus a threat to the purity of Israel's ancestral faith as represented in the wilderness period. Although the group survived for more than 250 years, it was likely never large in numbers.

The Old Testament as a whole seems to indicate that abstinence from the use of wine was exceptional rather than normative. The cases of total abstinence were few in number. More common were instances of temporary abstinence. Although certain dangers were recognized in its excessive use, there is no blanket condemnation of wine. Properly used, it was a legitimate part of normal Hebrew life.

2

The Teaching and Practice of Judaism

When we turn from the Old Testament to the noncanonical literature of Judaism, the same general positive appraisal of wine is continued. This attitude is reflected, for example, in the utterance of Ben-Sirach: "What is life to a man who is without wine? It has been created to make men glad" (Sirach 31:27b). Again, in the story of Ahikar, the sage says to his adopted son: "Son, God hath ordained wine for the sake of gladness" (Arm. Ver. 2:95). But Ben-Sirach is well aware of the dangers of excessive drinking and therefore strongly urges moderation.

> Do not aim to be valiant over wine,
>   for wine has destroyed many....
> Wine is like life to men,
>   if you drink it in moderation....
> Wine drunk in season and temperately
>   is rejoicing of heart and gladness of soul.
> Wine drunk to excess is bitterness of soul,
>   with provocation and stumbling (Sirach 31:25-29).

Considerable attention is given to wine in the Testament of Judah, chapters 14 and 16. In light of the lamentable consequences of drunkenness the patriarch strongly enjoins moderation. "Much discretion," he says, "needeth the man who drinketh wine . . . and herein is discretion in drinking wine, a man may drink so long as he preserveth modesty" (14:7). Indeed, he suggests, in order to avoid the evils of excess it might be better to avoid wine entirely (cf. 16:3).

In 3 Baruch reference is made to the Jewish tradition that the forbidden tree which led Adam astray was the vine. And "the men who now drink insatiably the wine which is begotten of it, transgress worse than Adam. . . . For those who drink it to surfeit do these things: neither does a brother pity his brother, nor a father his son, nor children their parents, but from the drinking of wine comes all evils such as murders, adulteries, fornication, perjuries, thefts and such like and nothing good is established by it" (4:16 f.).

Philo has a lengthy treatise on drunkenness. Not all of this essay has been preserved. In the part that has survived, Philo, although concentrating chiefly on the misuse of wine and the spiritual symbolism of such misuse, thought that it had a proper use. He observes that among those in the Pentateuch who used wine many are held "in the highest admiration for their virtue."[7]

The rabbis believed in this emphasis on moderation. They did not forbid the use of wine but regarded it as one of God's gifts to be received with thanksgiving.[8] Like the prophets, however, they condemned drunkenness. A decision issued by a judge when drunk was not valid. A drunken man's prayers were of no value. A priest was not permitted to drink wine before entering the temple to serve. Indeed, the efficacy of any religious act was nullified when the performer was intoxicated. Their objections were not only ceremonial but also moral. They were concerned about the bad moral consequences that usually follow in the wake of drunkenness. The lamentable conduct of Noah and Lot was regarded as illustrative of such evils.

Wine was freely used in Jewish life.[9] When taken in moderation it was regarded as promoting good health. A rabbi once suggested that for reasons of economy beer should be substituted for wine when possible. But his view was opposed on the ground that the preservation of health was of greater importance than economy. Wine also possessed medicinal value, and cures were ascribed to its use.

At least from Hellenistic times onward, wine was employed in the Passover Feast.[10] The special ritual celebrations that marked the beginning and the end of the Jewish Sabbath were observed with wine. Wine was served with "the meal of consolation" in the mourner's house on the occasion of death. Wine was served at wedding feasts and in late Judaism also at the meal celebrating the ceremony of circumcision. Indeed, for all religious ceremonies wine was regarded as preferable to other beverages. The demand for wine in Palestine in Roman times apparently was greater than the supply. At least there is evidence that wine was imported from the west. Many large Rhodian pottery wine jars found in excavated Palestinian cities bear witness to this once-flourishing business.[11]

This brief survey of the Old Testament and of Judaism may be concluded by a few summary statements which express the general attitude of this literature toward wine.

(1) Wine, as a part of God's creation, is inherently good. It can, therefore, legitimately be used by persons.

(2) Wine, like all other good gifts of God to people, can be abused. The excessive use of wine which results in drunkenness is such an abuse and is strongly condemned.

(3) The basis of such condemnation was not merely a pragmatic self or socal interest; it had deeper roots. It must be

viewed in the context of Israel's religious faith with its emphasis upon a person's relationship both to God and to others. Conduct which makes the fulfillment of those obligations impossible is not legitimate. This means that it will not do to account for the emphasis on moderation in Judaism merely as the result of Greek influence. In Judaism "the golden mean" was not an ultimate norm as in Greek thought. It was simply a helpful rule of thumb as the Jew sought to fulfill his destiny under God.

3

New Testament Teaching and Practice

The New Testament references to wine are not numerous. As might be expected, however, in light of what has already been said about Judaism, the Gospels do reflect a society in which wine was freely used. Jesus made reference to skin bottles used as receptacles for wine (Mark 2:22). The employment of wine for medicinal purposes is seen in the parable of the Good Samaritan (Luke 10:34). The practice of giving wine to which spice had been added as a drug to dull the pain of execution is reflected in the account of Jesus' crucifixion (Mark 15:23; Matthew 27:34). There is reference to the use of wine at wedding feasts (John 2:1-10) and to excessive drinking resulting in drunkenness (Matthew 24:49; Luke 21:34).

Jesus as a Jew appears to have followed traditional Jewish practice regarding the use of wine. He was not a Nazirite, as was John the Baptist. Indeed, He was accused by His enemies of being a notorious drinker (Matthew 11:19; Luke 7:34).[12] This was a slanderous remark, but it does indicate that Jesus did not abstain from the use of wine. A similar conclusion may be drawn from the account of the Last Supper. The Greek word for "wine" does not occur in the text. In its place is the phrase, "the fruit of the vine" (Mark 14:25). But this is merely a circumlocution for wine.[13]

Jesus' acceptance of wine as a legitimate beverage in Judaism is suggested also by the miracle He performed at the wedding feast in Cana (John 2:1-11). Probably in the intention of the author this miracle is to be understood as a Messianic sign pointing to the new order of Judaism. But to stress the symbolic significance of the story is not to empty it of historical meaning. Neither will it do to argue that what Jesus made on this occasion was simply unfermented grape juice. The word employed is the normal word for wine. Both classical Greek and the papyri employed another word for unfermented grape juice.[14] Even though we may wish it otherwise, honest exegesis compels the candid admission that on this occasion Jesus deliberately added to the stock of wine available for consumption at the wedding feast.[15]

If Jesus was not a teetotaler, He did, however, warn against drunkenness. In the apocalyptic discourse recorded in Luke 21 Jesus charged His hearers: "Take heed to yourselves lest your hearts be weighed down with dissipation and drunkenness and cares of this life, and that day come upon you suddenly like a snare" (v. 34). In the parable of the steward who abused his position and lived a riotous and drunken life, Jesus warned that such conduct would be met with severe punishment: "The master of that servant will come on a day when he does not expect him and at an hour he does not know, and will punish him, and put him with the unfaithful" (Luke 12:46).[16]

In the letters, as in the Gospels, the use of wine is not summarily forbidden. Indeed, its use is assumed. One of the qualifications for the office of deacon is that the candidate must not be "addicted to much wine" (1 Timothy 3:8). Likewise, an aspirant to the office of bishop must not be a drunkard (1 Timothy 3:3).[17] Titus is enjoined to bid the older women in the Christian community in Crete not to be "slaves to drink" (Titus 2:3). These directives do not forbid all use of wine, but they do seek to regulate its use. Curiously enough, there is one rather well-known passage in which its use is recommended. Timothy is advised: "No longer drink only water, but use a little wine for the sake of your stomach and your frequent ailments" (1 Timothy 5:23). The background of this exhortation is obscure. It is possible in light of 1 Timothy 4:1-5 that Timothy may have adopted a policy of total abstinence from wine in the belief that asceticism represented a higher standard of purity. Again, the suggestion has been made that since the advice is related to a practical regard for health, Timothy may have fallen prey to an insistence on spiritual healing without the use of the conventional remedies.[18] Wine was widely regarded both in Jewish and pagan circles as possessing medicinal properties.[19] Its help, therefore, in times of need should not be shunned. But whatever may have been the reason for Timothy's abstinence, he is urged to use "a little wine" for the sake of bodily health.

If the moderate use of wine is regarded as legitimate, drunkenness is repeatedly forbidden and condemned. Paul considers drunkenness as one of the works of the flesh which disqualifies a man for entrance into the kingdom of God (Galatians 5:21; 1 Corinthians 6:10). He is not slow to characterize the drunkenness which marked the Corinthian celebration of the Lord's Supper as shameful and invalidating the inner meaning of the observance (1 Corinthians 11:17-22). In the same letter he advised the members of the Corinthian church to refrain from intimate fellowship with any brother who was a drunkard (1 Corinthians 5:11). Paul admonished the Ephesian Christians not to seek inspiration in drunkenness and the dissipation that goes with it but to allow the Holy Spirit to fill them with true Christian exhilaration. They will thus be enabled to make a positive contribution to the edification of the Christian community (Ephesians 5:18). Writing to the Roman church Paul clearly indicates that drunkenness belongs to the world of darkness and does not befit the children of the day that is about to dawn in full eschatological splendor. They must, therefore, cast it aside and live as persons who belong to the new order (Romans 13:11-14).

It should be pointed out that none of these texts (nor any in the Old Testament), speaks to the matter of why persons drink. No attempt is made to offer psychological, physiological, or sociological analyses. The writers are content to deal with overt conduct and to evaluate it accordingly.

This then concludes the survey of explicit references in the New Testament to the use of wine. These materials point to its casual acceptance as a common beverage of the day while condemning its excessive use which leads to drunkenness. This position might be summed up as one of moderation. It is essentially the same attitude as found in the Old Testament and in Judaism. Indeed, the support for such a stance is not lacking even outside the Christian community in the Greco-Roman world. The Stoic philosophers stressed temperance and self-control in the matter of drinking as in other activities. Seneca has a long tirade against the vice of drunkenness. Galen, one of the most famous physicians of the Greco-Roman world, while recognizing the medicinal value of wine, condemns its immoderate use.[20] It would be grossly unfair to the New Testament, however, to conclude that its norm is no different from that of the contemporary world of its day. Likewise, it would not be dealing adequately with the New Testament to conclude that its implicit and explicit approval of wine in moderation provides a clear and normative answer to the problem in our society.

Such a direct transition from the New Testament to the modern scene overlooks significant differences between the two situations. The wine of the New Testament was not distilled liquor but was made by the process of natural fermentation. The alcoholic content probably did not exceed 5 to 8 percent. The potency was further weakened by the practice of mixing wine and water.[21] Also, the wine was drunk mostly in connection with meals. The modern tavern was unknown in Palestine. Moreover, on those occasions when drunkenness did occur, the disastrous social consequences were negligible in comparison with those in the highly mechanized modern society of our day.

But these considerations, important as they are, do not go to the heart of the matter. We have been occupied with the specific details of the biblical text. This is basic and highly important. But to gain a correct perspective we need to stand back a bit from the specific material dealing with the problem under study and attempt to look at the New Testament ethic as a whole. To change the imagery, consider C. Anderson Scott's suggestive analogy from the world of sports.

> All games which deserve the name are provided with laws. And the first condition of playing the game is that we know the laws and obey them. But all the best games are provided with something more. That more may be described as the etiquette of the game. And the etiquette is even more important than the laws.... So life has its laws; but it has also its etiquette. And many things which are not forbidden by the laws are sternly banned by the etiquette.[22]

This distinction between laws and etiquette suggests our further task. What is the etiquette or style of the New Testament ethic? What is its general character, its frame of reference, its basic motivation, and its central thrust? Not until we have gained some understanding of these matters shall we be prepared to arrive at valid conclusions in our study. Let us briefly sketch some of the main outlines of the New Testament ethic as they bear upon our present problem.

4

New Testament Ethical Perspectives

First, the ethic of the New Testament has a religious rather than a philosophical basis. Emil Brunner has defined Christian ethics as "the science of human conduct as it is determined by divine conduct."[23] This means that human conduct is not anchored in naturalistic understandings of man or society. Its ultimate norms are not found in such considerations as the happiness of the individual or the greatest good for the greatest number. Unlike Greek ethics which sought their foundations in human reason, biblical morality is shaped by a word from beyond us. Biblical ethics are the product of biblical religion. They may properly be called "response" ethics. Severed from their theological rootage they wither and die. The "ought" of human duty is firmly grounded in the "is" of what God has already done.

This may be seen in the Old Testament. It should never be forgotten that the Decalogue is set in the context of God's prior action of grace (Exodus 19 and 20). God had redeemed His people from Egypt. The law was an attempt to give structure to

Israel's grateful response to God's grace. Similarly, in the New Testament the Sermon on the Mount is prefaced by the declaration that "the kingdom of heaven is at hand" (Matthew 4:17; cf. v. 23). The Beatitudes which inaugurate the sermon are not words of demand but affirmations of grace. In the letters of Paul ethical instruction is regularly preceded by theological exposition. The imperative grows out of the indicative. Since God has already acted, we now must act.

But more is involved than the matter of motivation when the New Testament grounds ethics in redemptive history. The advent of Christ and the inauguration of a new order of life provide a frame of reference for ethical action that must be noted. Conduct must be oriented to the revelation of God's pattern for life and God's destiny for us in Jesus Christ. Thus Paul, for example, can employ terms which were also used in Greek ethics. But they are given a new significance when they are set in the Christian context. Writing to the Ephesians he says there are certain practices which must not be found among them because they are not in accord with what "is fitting among saints" (Ephesians 5:3 f.).[24] What is the norm in view? It is nothing less than their Christian status and destiny as disclosed in the Gospel. Paul uses such terms as *conscience, mind, law,* and *the good,* all of which are significant for ethics. But their meaning is no longer determined by the premises of Stoicism but in relationship to Jesus Christ. Conduct must be shaped in the light of His character and will, rather than by reference to the conventional norms of sub-Christian culture. It is not the light of nature but the grace of God that has appeared in Christ that "teaches us to have no more to do with Godlessness or the desires of this world but to live, here and now, responsible, honorable and God-fearing lives" (Titus 2:12, Phillips).[25] We are summoned to live in a manner "worthy of the gospel of Christ" (Philippians 1:27). Here, then, is the touchstone by which the legitimacy of any act must be tested.

Second, the ethic of the New Testament is nonlegal in character. Specific injunctions abound in the pages of the Gospels and the letters. This is understandable in light of the fact that the New Testament is not an abstract philosophical discussion but an "occasional" work. Its message was addressed to specific persons and definite situations. It reflects the realism of given historical occasions. It would be a mistake, however, to regard these pointed instructions as constituting a type of code book for Christian conduct similar to the Mishnah or the Talmud in Judaism. The ethical materials in the New Testament are far removed in spirit from the legalistic, casuistic contents of the Jewish law books. They resemble more closely the prophetic approach to ethics. They are pervaded by a vitality and freshness that is not found in a code book. They represent deep and far-ranging principles. They point to the ideals and goals that ought to govern the life of individuals in the new order of grace. They resemble more closely a compass than a road map. They serve to give direction to life rather than directions.[26]

This may be illustrated in various ways. Jesus, for example, in Matthew 5:21-48 is combating an inadequate understanding of the Mosaic law which gave a too restricted meaning to the precepts of the Pentateuch. He sought to get hold of the basic underlying intention of God for human life which Pharisaic literalism had obscured. He summed up the whole law in the two commandments which enjoin love for God and for one's fellowmen (Mark 12:28-31). Such a norm is incompatible with a petty legalism. It is at once both more comprehensive and more demanding than a code-book morality.

Paul, likewise, finds in the law of love the supreme norm of Christian conduct. Where love controls life, that which the law in its multitudinous precepts aims at is fulfilled (Romans 13:8-

10). He speaks of life in Christ as freedom from the "yoke of slavery," in which, "led by the Spirit," we are no longer "under the law" (Galatians 5:1, 18). He summons his readers to possess "the mind of Christ" as they deal with the practical problems of daily conduct (Philippians 2:5). This is a call to a perspective which is dynamic and creative in its approach to ethics.

To affirm that the New Testament ethic is nonlegal in character is not to suggest that it is without specific shape or form. If Christian love is the ultimate principle of Christian action, the meaning of love is not left ambiguous. Here is where the specific injunctions of the New Testament fall into place. They are designed to make clear something of the meaning of love in life. They are concrete illustrations of what life under the control of the gospel ought to look like. They are "landmarks along the way" which "reassure us when we have read the compass aright and they trouble us when we twist the reading of that compass to our own advantage."[27]

If the New Testament is not a code book offering detailed answers to all eventualities of Christian conduct in dealing with specific personal or social problems, then we must seek to find the direction in which it points. We must attempt to ascertain the spirit and intent of what it says rather than be content with a bald literalism which may deny the very spirit of the gospel. This may mean that the ultimate outworking of the Christian ethic may take us beyond what is expressly stated in the New Testament or what was practiced in the early church.

A ready example of this is the matter of slavery. Slavery, as such, is not forthrightly condemned anywhere in the New Testament. Yet it is clear that the gospel which Paul brought to bear upon the problem as it existed in Colossae would, when it had borne its full fruit, demolish the institution. What, then, are the implications of the gospel for dealing with the problem of the use of alcoholic drink? To answer this question, it is necessary to pay attention not only to what is said but also to what is implied.

Third, the intent of the New Testament ethic is the enrichment and fulfillment of human life in the highest sense. Deep in the human soul there is a thirst for life. The gospel speaks to this desire of our hearts. Jesus declared that He had come that persons "may have life . . . and have it in all its fullness" (John 10:10, NEB).[28] He proclaimed the advent of the kingdom of God, and the inner meaning of that kingdom could be defined as entrance into life (Mark 9:45, 47). The message which the apostles were commanded to preach is described as "the words of this Life" (Acts 5:20). The author of 1 John 5:12 could sum up the meaning of the Christian life in a single sentence, doubtless out of his own experience: "He who has the Son has life; he who has not the Son of God has not life." It is not surprising then that some early Christians who were condemned to work in the salt mines of North Africa are said to have cut with their instruments of labor on the walls of the mines the words *vita, vita, vita* (life, life, life). They had caught the meaning of the gospel.

The Christian ethic must be understood in the light of this positive thrust of the gospel. Discipleship is not a scheme designed to impoverish life but the path to its true fulfillment. The Christian ethic is not meant to dry up the springs of true joy but to put a song in the heart and a light in the eye. The New Testament, like the Old, places no premium upon asceticism per se, as though it represented a superior sanctity. The created order and our normal involvement in it are not inherently evil. They are God's good gifts to us designed to contribute to our welfare and happiness. Thus the effort, due to Hellenistic influence, to prohibit marriage and the eating of certain foods as sinful in themselves is condemned as a departure from the faith (1 Timothy 4:1-5; cf. Colossians 2:16, 20 ff.).

Salvation is not escape from normal life in the world but its redemption and enrichment.

But the New Testament also recognizes the possibility of abusing God's gifts and of employing them to wrong ends. Many of the vices so vigorously denounced are simply perversions of corresponding and desirable virtues. We may thwart God's purpose for us by our careless living. Life can reach its intended maximum only when God's pattern for it is embraced and those disciplines necessary to achieve it are voluntarily accepted. This is the rationale for whatever negative or "ascetic" notes occur in the New Testament. Whether we like it or not, we have been made for fellowship with God. Whatever conduct militates against the realization of this great purpose in our lives must resolutely be laid aside. Whatever impairs the mind or the body which is the temple of the Holy Spirit cannot be tolerated. Whatever "steams the windows of the soul" so that we cannot see God nor reflect His glory in our lives, must be shunned. Whatever threatens to overpower us and take us into captivity so that we cannot serve God with complete abandon, must be judged unworthy of a place in our lives.

Fourth, the New Testament ethic is concerned with the conduct of the individual not only in its personal reference but also in its social effect. This interest is not missing from the teaching of Jesus (e.g., Matthew 18:5 ff.), but it is most fully developed by Paul. In his letters it is clear that although "we become related to Christ singly ... we cannot live in Christ solitarily."[29] The Christian is a member of the church which Paul repeatedly describes under the imagery of the body of Christ. This figure among other things vividly portrays mutual dependence and interaction in the Christian fellowship. This is developed at some length in 1 Corinthians 12:12-30. This corporate conception, however, is in the background of all that Paul has to say about the life of the individual Christian.

Its ethical implications are drawn out in two extended passages dealing with the matter of Christian liberty: Romans 13—15 and 1 Corinthians 8—10. The basic concept in these passages is that of community under the lordship of Christ. While the Christian freedom of the individual is not ignored, it is set in the context of the welfare of the church. "The limit of liberty," wrote Percy Gardner, "is not a rule, however reasonable, but an enthusiasm. Love makes liberty stop of her own accord and willingly give up the extreme of her course."[30] Indeed, the emphasis here is not upon personal rights but upon social obligations.

Christian freedom, then, is not the opportunity to do as one wishes but the freedom to do as one ought. Personal duty involves responsibility for our brother's and sister's well-being. When the Corinthians proudly boasted, "All things are lawful," Paul reminded them that "not all things build up." Paul had in mind group edification, as is clear from what follows: "Let no one seek his own good, but the good of his neighbor" (1 Corinthians 10:23 f.). Now it is precisely with this thought in mind that Paul has a word to say about the use of wine. The Christian must act in such a way that he will not place

> ... a stumbling block or hindrance in the way of a brother.... If your brother is being injured by what you eat, you are no longer walking in love. Do not let what you eat cause the ruin of one for whom Christ died.... It is right not to eat meat or drink wine or do anything that makes your brother stumble (Romans 14:13, 15, 21).

The application of the principle of social responsibility may be extended, of course, beyond the specific formulation of it in Romans 14. There Paul has in view the influence of personal example upon the conduct of a fellow Christian. This is also the primary concern in 1 Corinthians 8—10. However, in this latter passage Paul's thought ranges beyond the confines of the Christian community to include the non-Christian. His final word in this treatment is sweeping in its scope.

> So, whether you eat or drink, or whatever you do, do all to the glory of God. Give no offense to Jews or to Greeks or to the church of God, just as I try to please all men in everything I do, not seeking my own advantage, but that of many, that they may be saved (1 Corinthians 10:31-33).

Reference may be made in this connection to what was said earlier regarding the difference in the complexity of society today as compared to that of the first century. The potential for causing widespread suffering and destruction through the misuse of one's freedom has been vastly compounded in our day by modern technology. The horse-drawn chariot was the fastest vehicle in the ancient world. Even if human sense were impaired by the use of too much wine, horse sense conceivably might take over and avert serious consequences. But the situation is otherwise if a mind befuddled by alcohol is behind the controls of a high-powered automobile on the highway or a jet airplane loaded to capacity descending to a runway or before a radar screen attempting to determine whether weapons of mass obliteration should be unleashed against mankind. Present-day science does not increase the truth of the biblical obligation that we are our brother's and sister's keeper, but it does underscore the seriousness of its neglect.

5

Conclusion

We have briefly surveyed the biblical practice and teaching regarding the use of wine. To what conclusion has this study brought us?

It is clear that wine was freely used in the Hebraic-Judaic-Christian community. Although some individuals abstained for varying periods of time for special religious reasons, general abstinence was neither taught nor practiced in biblical times. There is a decided emphasis, however, on moderation in usage. Drunkenness is consistently and severely condemned throughout Scripture. We noted, too, that the wine of which the Bible speaks was made by the process of natural fermentation and was relatively low in alcoholic content. There is evidence, also, that wine was diluted with water in the Greco-Roman period. Furthermore, the wine normally was drunk in connection with meals and not in public bars or drinking parlors.

Does the Bible, therefore, sanction the use of alcoholic beverages for Christians today? There is no explicit answer to this question in the Bible. This does not mean, however, that it sheds no light on the problem. The answer must be derived from the general thrust of the biblical message. If we would not abuse the Bible, we must set all that modern science and observation has taught us about the physical and social effects of the use of alcohol in the light of the revelation of God's will and purpose for us in the gospel. We dare not adopt uncritically the norm of modern society on this matter. We are ultimately responsible to the demands of Christ upon us. In all seriousness, then, we must seek to determine the implications of Christian discipleship regarding the use of alcoholic beverages. Although not all Christians feel compelled to adopt a policy of abstinence, strong and valid arguments support this position.

First, contrary to popular opinion, alcohol is not a stimulant but a depressant. It affects the body as any narcotic or anesthetic would. It removes inhibitions and enables a person to say and do things which normally would not be thought proper. It impairs judgment and slows up reactions. It blunts the memory and permits a false self-confidence to emerge. It offers an escape from the realities of life which may look attractive at the moment but in the end is no satisfactory solution. The opportunities and responsibilities of life are both too exciting and too demanding to take into one's body beverages which tend to decrease rather than increase the efficiency of mind and body to meet them.

Second, even though the undesirable effects of alcoholic beverages become evident only when indulgence has passed beyond limited use, a hidden danger nevertheless lurks in what is often called "moderate" drinking. It is the possibility of becoming a slave to alcohol. The causes for this condition are complex. But it should not be forgotten that there is no way of predicting accurately who among those who begin to drink will eventually become an alcoholic. It is estimated that the chance of this occurring may be as high as one in ten.

The matter might be put more graphically by the use of an analogy. Imagine a person going to an airline terminal to purchase a ticket.

When the ticket is issued the agent says: "You should be aware of the fact, sir, that on this plane, seating one hundred persons, ten seats sometime during the flight will suddenly give way and drop their occupants out of the plane."

With understandable concern the traveler responds: "Oh, sir, please see to it that I am not placed in one of those seats!"

But the agent replies: "I'm sorry but we don't know in advance which seats will give way. Have a good flight, sir!"[31] It is safe to say that not many persons would be willing to take that sort of a gamble. No alcoholic ever intended to become one. The only certain guarantee of not becoming an alcoholic is to adopt a policy of abstinence.

Third, even if the use of alcoholic beverages could personally be kept within the bounds of moderation, there is still another important dimension to the problem which the Christian cannot ignore. The suffering, slavery, and ruin which alcohol can produce in human life and society are well known, as well as the fact that some persons seem to be disposed for physical or psychological reasons to the danger of an uncontrollable addiction to alcohol once the habit has been established. In light of this, should not the Christian wish by his own example to make abstinence for his brother an easier task rather than more difficult? Social pressure is a powerful determinant of conduct. In these days when drinking is widely accepted as a social practice in society, abstinence frequently demands more courage than some persons of themselves seem able to command. As Christians we must seriously ask whether it is not our duty and privilege to lend their faint courage encouragement by a personal example of abstinence. Should we not make it less rather than more difficult for others to avoid being ensnared by drink?

Fourth, one further word should be added which in a sense is the presupposition of the statements thus far made. Christian freedom regarding the use of alcoholic beverages should not be thought of primarily, as is frequently the case, in terms of the right to drink. There is another and more exciting way to view it. It is the freedom not to drink.

There are many reasons why persons who drink do so in varying amounts and in a variety of circumstances. Among the reasons most often cited are to relieve tension, to promote sociability, to bolster one's self-image, and to conform to social expectations. These, and other motives, could be explored at length.[32] They pinpoint general problems which affect most of us at one time or another.

Anxieties and tensions are common experiences in life. The question, however, is whether reliance upon alcohol as a way of coping may not only prevent a frank confrontation with the underlying causes but also make us less able to deal successfully with them.

The desire for warm and meaningful relationships with others is a legitimate urge. The question, however, is whether the quality of relating is enhanced or impaired when mental alertness and good judgment are dulled through dependence upon alcohol.

The need for a positive self-image is essential to a wholesome self-respect. The question, however, is whether such a self-understanding to have integrity does not need a more secure foundation than the temporary lift provided by alcohol.

The wish for social approval is deeply rooted in human nature. The question, however, is whether the effort to fulfill this need can afford to ignore or override other considerations regarding the use of alcohol.

Are not sufficient resources available within the Christian faith and in the Christian community to enable Christians to work at their problems without needing to resort to the "aid" that alcoholic beverages "promise"? Many have found this to be true. For them abstinence is not so much a matter of a conscious restraint of their freedom as it is the spontaneous product of a life that is deeply rooted in Christ and is daily sustained by His grace and power known in and through the Christian community.

Notes

1. The figures cited are estimates. They have been drawn mainly from "The Third Special Report to the U.S. Congress on Alcohol and Health" (June 1978) by the Secretary of Health, Education, and Welfare, Joseph A. Califano, and from an interview with the Secretary by Jay Lewis reported in *The Journal of Studies on Alcohol*, 40.5, May 1979, pp. 539-543. For additional statistical analyses see Robert L. Hammond, *Almost All You Ever Wanted to Know About Alcohol but Didn't Know Who to Ask*, copyright 1978 by the American Business Men's Research Foundation and published by Michigan Alcohol and Drug Information Foundation, Lansing, Mich.

2. E.g., Proverbs 23:29-35 and 1 Timothy 5:23.

3. See "The Story of Sinuhe," *The Ancient Near East: An Anthology of Texts and Pictures*, ed. by J. B. Pritchard, Princeton University Press, Princeton, 1958, pp. 5 ff.

4. Sirach. 50:15; cf. Josephus, *Antiquities*, III. ix 4.

5. See W. F. Albright, *The Archaeology of Palestine*, rev. ed., Penguin Books, Inc., Baltimore, 1950, p. 115.

6. G. von Rad, *Old Testament Theology*, trans. by D. M. G. Stalker, Vol. I, Oliver and Boyd, Edinburgh, 1962, p. 63.

7. *De Ebrietate*. See esp. I, 1-4.

8. There are many references to wine in rabbinic literature. A selection is conveniently available in the article "Wine" by J. E. Eisenstein in the *Jewish Encyclopedia*, XII, pp. 582 ff. Cf. also I. W. Raymond, *The Teaching of the Early Church on the Use of Wine and Strong Drink*, Columbia University Press, New York, 1927, pp. 45 ff.

9. Water was ordinarily drunk at meals except on festal occasions when wine was the proper beverage. For references see J. Jeremias, *The Eucharistic Words of Jesus*, 3rd ed. trans. by Norman Perrin, S.C.M., London, 1966, pp. 50 ff.

10. The first reference to such use is Jubilees 49:6.

11. W. F. Albright, *op. cit.*, p. 152.

12. The Greek *oinopotas* is translated by the RSV "drunkard" and the NEB "drinker." The sense would seem to be one given to the use of much wine.

13. See J. Jeremias, *op. cit.*, p. 183 f. The question whether the Last Supper was a true Passover celebration or a *Kiddush* need not concern us, since wine was used on both occasions.

14. The word for wine is *oinos*, while that for unfermented grape juice is *trux*.

15. For attempts to prove that the "wine" at Cana was unfermented grape juice, see F. R. Lees and D. Burns, *The Temperance Bible Commentary*, 5th ed., S. W. Partridge, London, 1880, pp. 304-306; also J. P. Free, *Archaeology and Bible History*, 5th ed., rev., Scripture Press, Wheaton, 1956, pp. 354 f. The argument, however, is based upon assumptions which are not supported by evidence.

16. The references to drunkenness in the Gospels are meager. Probably in Palestine in Jesus' day the vice was mainly confined to the wealthy class and was not common among the people with whom Jesus mainly associated.

17. Cf. Titus 1:7. The Greek term *paroinos* used in both passages is translated "given to drink" in the NEB. Excessive drinking is in view. The use of the word "temperate" *(naphalios)* in 1 Timothy 3:11 and Titus 2:2 probably also has in view moderation in the use of wine.

18. F. D. Gealy in *The Interpreter's Bible*, ed. by G. A. Buttrick, Abingdon Press, Nashville, 1955, *ad. loc.*

19. Note the rabbinical statement: "Wine is the greatest of all medicines; where wine is lacking there drugs are necessary" (B.B. 58b.). Plutarch prescribed wine to those needing a tonic and recommended it especially for stomach ailments *(Sympos.* iii. 5. 2). Cf. Strabo, *Geog.*, xiv. 1. 15.

20. For this and the preceding reference see I. W. Raymond, *op. cit.*, pp. 74 ff.

21. Cf. 2 Maccabees 15:39.

22. *New Testament Ethics*, Cambridge University Press, New York, 1930, p. 71.

23. *The Divine Imperative*, trans. by Olive Wyon, Westminster Press, Philadelphia, 1947, p. 86.

24. Cf. also 1 Timothy 2:10 for another occurrence of the same term.

25. From *The New Testament in Modern English*, © J. B. Phillips, 1958. Used by permission of The Macmillan Company and Geoffrey Bles Ltd.

26. Cf. T. W. Manson, *The Sayings of Jesus*, S.C.M. Press, London, 1949, pp. 35-38.

27. H. K. McArthur, *Understanding the Sermon on the Mount*, Harper and Brothers, New York, 1960, p. 160.

28. © The Delegates of the Oxford University Press and the Syndics of the Cambridge University Press 1961.

29. J. A. Mackay, *God's Order: The Ephesian Letter and This Present Time*, Macmillan, New York, 1953, p. 117.

30. *The Religious Experience of St. Paul*, Williams and Norgate, London, 1911, pp. 35 f.

31. The substance of this illustration is drawn from the editorial, "Abstinence Makes Sense," *Christianity Today*, Vol. VIII (April 24, 1964), p. 25. The proportionate figures, however, are based on studies of the National Institute on Alcohol Abuse and Alcoholism which estimates that one in ten of the Americans who now drink is either a full-fledged alcoholic or at least a problem drinker. See B. Q. Hafen, *Alcohol: The Crutch That Cripples*, West Publishing Company, St. Paul, Minn., 1977, p. 47.

32. Helpful analyses are available in the following: William McCord and Joan McCord with Jon Gudeman, *Origins of Alcoholism*, Stanford University Press, 1960; Howard J. Clinebell, Jr., *Understanding and Counseling the Alcoholic*, Abingdon Press, 1968; John Jung, "Drinking Motives and Behavior in Social Drinkers," *Journal of Studies on Alcohol*, 38.5, May 1977, pp. 944-952; G. E. Barnes, "The Alcoholic Personality: A Reanalysis of the Literature," *ibid.*, 40.7, July 1979, pp. 571-634.

Howard H. Charles was born at Lititz, Pennsylvania. He was educated at Eastern Mennonite College, Goshen College, Union Theological Seminary (Richmond, Va.), Princeton Theological Seminary, and the University of Edinburgh, Scotland. He is married to the former Miriam Stalter of Elida, Ohio, and is the father of two sons.

He served as a pastor for four years at the Lititz Mennonite Church, Lititz, Pa. He began teaching in the Goshen Biblical Seminary in 1947. More recently the school has been relocated in Elkhart, Indiana, where he is presently professor of New Testament in the Associated Mennonite Biblical Seminaries.

In 1961 he was in Japan on an assignment under the Mennonite Board of Missions as a Bible teacher in the Japanese Mennonite Church. He served in a similar capacity in Ghana in 1973-74. Since 1950 he has been a regular contributor to the Herald Press Sunday school literature.